SRA

Connecting Math Concepts

Level A Workbook 2

COMPREHENSIVE EDITION

A DIRECT INSTRUCTION PROGRAM

McGraw Hill Education

Bothell, WA • Chicago, IL • Columbus, OH • New York, NY

mheducation.com/prek-12

Send all inquiries to:
McGraw-Hill Education
4400 Easton Commons
Columbus, OH 43219

ISBN: 978-0-02-103573-1
MHID: 0-02-103573-3

Printed in the United States of America.

20 LON 25 24 23

Name ____________________

$\square$ $\square$

$7+3=\square$

$2+6=\square$

$7-3=\square$

$6+\square=$

|||||

$18+\square=$

||||||

7 5

6 9 6

8 3

19+1=	6+1=	4+1=
8+1=	15+1=	7+1=
32+1=	3+1=	9+1=

□ −7=

□ −4=

16+5=

9+4=

Name ______________________

 $6 - 4 = \square$

$8 + 3 = \square$

$7 - 2 = \square$

 $\square$

$\square$

 ______________________

____

 $14 + \square =$

$23 + \square =$

14

17

26

☐ − 5 =

|||||||||||

☐ − 3 =

|||

9 + 1 =	13 + 1 =	10 + 1 =
15 + 1 =	2 + 1 =	19 + 1 =
11 + 1 =	27 + 1 =	8 + 1 =

17 + 5 = ☐

8 + 6 = ☐

Name ____________________

____________________ ____________________

$7 - 3 = \square$

$6 - 5 = \square$

$24 + 3 = \square$

25 + ☐ =

||||||

27 + ☐ =

|||||

85

19

☐ − 7 =

|||||||||

☐ − 4 =

||||||||||||

14 + 1 =

7 + 1 =

19 + 1 =

3 + 1 =

9 + 1 =

25 + 1 =

8 + 1 =

13 + 1 =

16 + 1 =

Name ______________________________

$16 + \square = 23$

$11 - 4 = \square$

$35 + \square = 39$

$4 - 3 = \square$

$6 + 5 = \square$

$7 - 1 = \square$

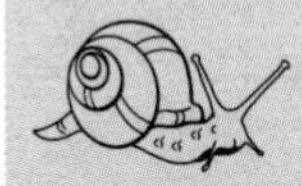

$\square$ $\square$

 ☐ ☐ ☐ ☐

1 1

7 2

3 8

9 + ☐ =

||||||||

10 + ☐ =

||||||

 ☐ − 6 =

||||||||||

☐ − 2 =

|||||||||||

17+1=	21+1=	2+1=
14+1=	8+1=	16+1=
9+1=	5+1=	3+1=

Name ____________________

$8 + 5 = \square$

$7 - 6 = \square$

$32 + 6 = \square$

$4 + \square = 9$

|||||

$6 + \square = 10$

||||

$14 + \square = 17$

|||

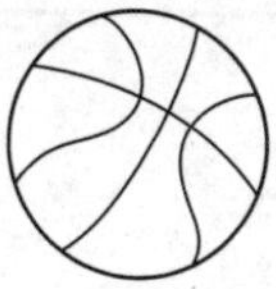

$\square$ $\square$ $\square$ $\square$

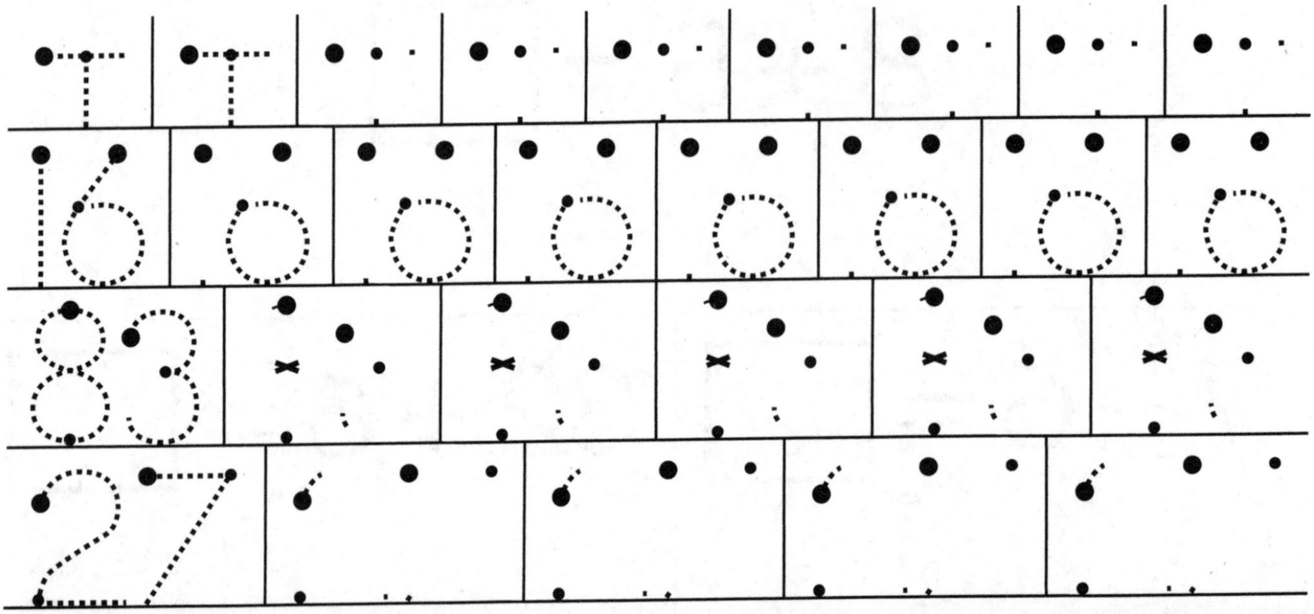

8 + □ =
|||||

13 + □ =
||||||

□ − 5 =
|||||||

□ − 3 =
||||||||||||

9+1=	12+1=	16+1=
8+1=	27+1=	14+1=
10+1=	21+1=	25+1=

Name ______________________________

______________ | ______________

TTTT
TT

TTTT
TTTT

29 + ☐ = 31
||

14 + ☐ = 18
||||

☐

☐

34 + 5 = ☐

6 − 3 = ☐

1

58

40

8 − 3 =

16 + 5 =

5 − 4 =

32 + 7 =

□ − 7 =

|||||||

□ − 4 =

||||||||||||

14 + 1 =

9 + 1 =

5 + 1 =

18 + 1 =

11 + 1 =

7 + 1 =

13 + 1 =

2 + 1 =

10 + 1 =

$14 + \square = 17$ | $25 + \square = 29$

□
T T T T
T T T T

□
T T T T T

______________________ | ______________________

30
T T T T
T T T

60
T T T T
T T T

10
T T T T
T T T

63

25

6 + □ =
|||||||

18 + □ =
||||||

□ − 5 =
|||||||||||

□ − 4 =
||||||

7 + 5 =

7 − 6 =

8 − 2 =

9 + 7 =

9 + 1 =

35 + 1 =

17 + 1 =

24 + 1 =

13 + 1 =

28 + 1 =

Name ______________________________

$8 + \square = 12$

$41 + \square = 46$

$\square$

TTTTT
TTTT

$\square$

| | | | | | |

$\square$

TTTT

$\square$

| | | | | |

20

50

98

36

24

$36 + \square =$ ||||||

$7 + \square =$ |||||||

$\square - 8 =$ ||||||||

$\square - 3 =$ |||||||||||||

$5 - 4 =$

$7 + 3 =$

$6 - 3 =$

$48 + 4 =$

$36 + 1 =$

$9 + 1 =$

$23 + 1 =$

$12 + 1 =$

$21 + 1 =$

$15 + 1 =$

Name ______________________________

7 + ☐ = 12 | 36 + ☐ = 40

40 | 10

______________ | ______________

☐
IIIIIII

☐
TTT

☐
TTTT
TTTT

75					
20					
86					

23 + 4 =	5 − 3 =
8 − 1 =	6 + 5 =

8 + □ = IIIII	14 + □ = IIIIII
□ − 3 = IIIIIIIIII	□ − 5 = IIIIIIIII

9+1=	26+1=	14+1=
37+1=	5+1=	18+1=

______________ ______________

☐
T T T T T T

☐
| | | |

☐
T T T T T
T T T T

☐
| | | | | | | | | | |

15 + ☐ = 18

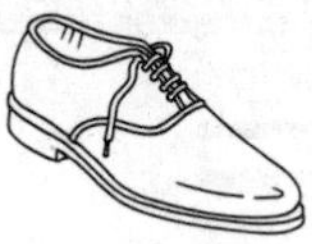

30

50

94

67

23

32 + 5 =

10 − 6 =

7 + 4 =

9 − 2 =

74 + □ =
||||||

31 + □ =
||||||||

□ − 5 =
||||||||||||

□ − 3 =
||||||||

45 + 1 =

21 + 1 =

13 + 1 =

8 + 1 =

37 + 1 =

9 + 1 =

Name ______________________

$6 + 0 =$

$32 + 0 =$

$9 + 0 =$

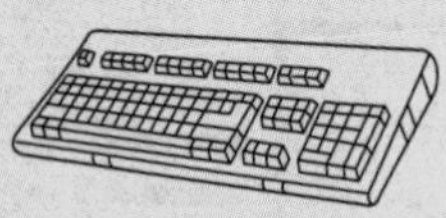 ______________________

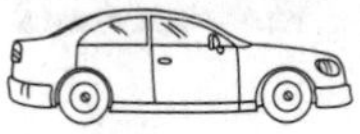

. 40

. 7

. 30

$34 + \square = 40$

$3 + \square = 8$

$\square$

$\square$

$\square$

96					
38					
72					

$9 + \square =$ (||||||)

$74 + \square =$ (|||||)

$\square - 6 =$ (|||||||||||)

$\square - 2 =$ (||)

$35 + 4 =$

$8 - 2 =$

$6 + 5 =$

$10 - 3 =$

$12 + 1 =$

$9 + 1 =$

$43 + 1 =$

$16 + 1 =$

$34 + 1 =$

$27 + 1 =$

Name ____________________

$18 + 0 =$

$24 + 0 =$

$0 + 0 =$

____

. 50

. 6

$8 + \square = 13$

$41 + \square = 45$

□

||||||||||

□

TTTT
TTT

14

50

23

□ − 8 =
||||||||

□ − 1 =
||||||||||||

35 + □ =
||||

17 + □ =
||||

6 − 5 =

8 + 4 =

8 − 3 =

13 + 5 =

4+1=	26+1=	18+1=
9+1=	30+1=	43+1=

Name ______________________________

15+1=

8+1=

42+0=

19+1=

34+0=

16+□=19

37+□=42

=

=

=

□

TTTT IIIII

□

TTT IIIIIII

3 .

60 .

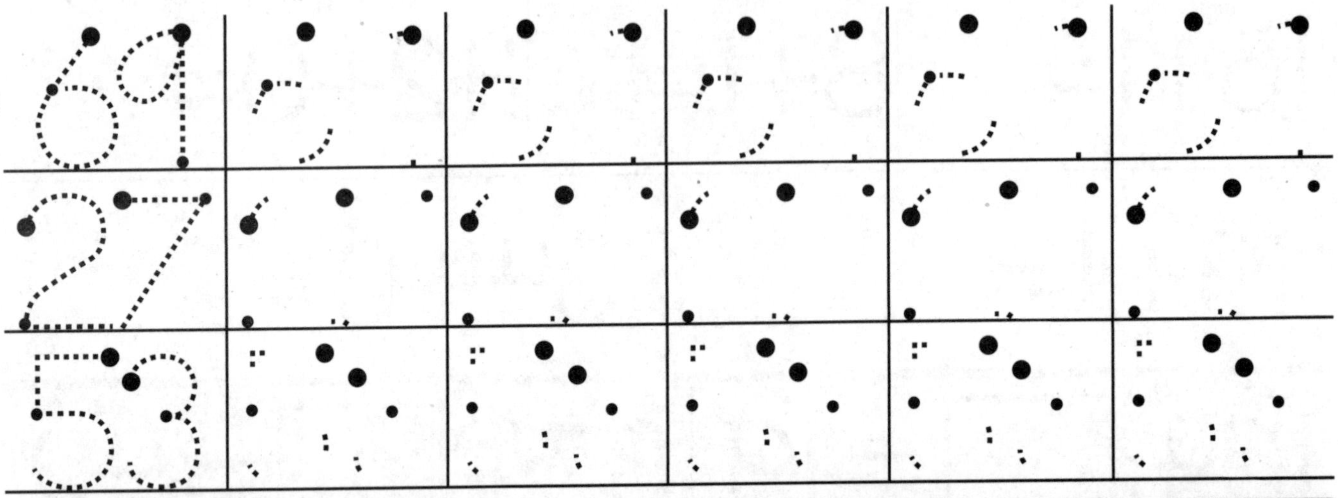

$\square - 4 =$

|||||||||

$\square - 6 =$

||||||||||

$13 + \square =$

|||||||

$7 + \square =$

||||

$7 - 5 =$

$6 - 2 =$

$12 + 4 =$

$23 + 5 =$

Name ____________________

☐
TTTIIIII

☐
TTTTIII

73 + ☐ = 78 | 9 + ☐ = 15

3+1= | 17+0= | 78+1=

12+1= | 56+0=

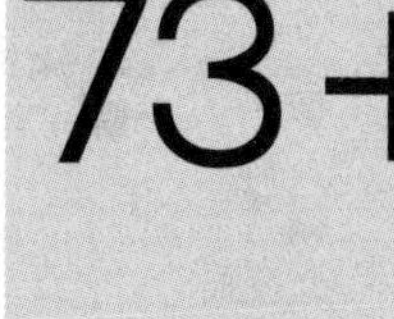

____

=

=

8 | 40 | 7 | 30

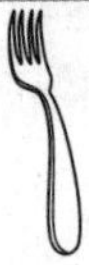

8 4 3

$\square - 10 =$

$78 + \square =$

$16 + \square =$

$\square - 4 =$

$39 + 3 =$

$9 - 4 =$

Name ______________________________

6 − 1 = ☐

6 − 0 = ☐

4 − 1 = ☐

10 − 0 = ☐

____

____

=

=

16 + ☐ = 20

48 + ☐ = 53

☐

T IIIIIII

☐

TTTT IIIIII

72 + 0 =	40 + 1 =
19 + 1 =	88 + 0 =

♡ 50	8	30	6

5

96

83

□ − 7 =	63 + □ =
|||||||||	||||||||

54 + 3 =	7 − 6 =
5 − 2 =	8 + 5 =

Name ____________________

$24 + 1 =$

$17 - 0 =$

$38 + 1 =$

$55 + 0 =$

□ + □ = 42

□ + □ = 37

____

□
TTTIIIII

□
TTTTTIII

34 + □ = 40

 60 | 5 | 9 | 20

2 4

5 3

8 9

73 + ☐ =
||||||||

☐ − 3 =
||||||||||

8 − 5 =

45 + 6 =

13 + 1 = | 19 + 1 = | 5 + 1 =

28 + 1 = | 6 + 1 = | 9 + 1 =

Name ____________________

 =

 =

 ____________________

□

□

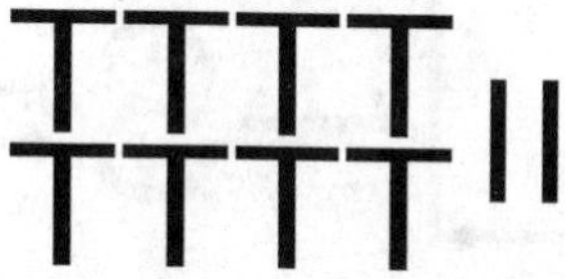

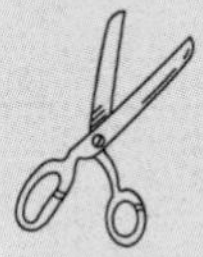 □ + □ = 51

□ + □ = 27

21 − 0 =

46 + 1 =

13 + 1 =

99 + 0 =

57 + □ = 61

39 + □ = 41

8	50	7	30

100

96

□ − 3 =

||||||||||

16 + □ =

||||||||

73 + □ =

||||||

□ − 9 =

|||||||||

8 − 5 =

19 + 5 =

Name ____________________

$3+5=\square$

$2+\square=6$

=

=

. 34

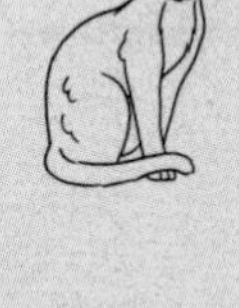

$\square+\square=74$

$\square+\square=38$

$\square$

TTT IIIIII
TT

$\square$

T IIIIIIII

$\square$

TTTT IIII
TTT

98

25

54 + 6 =	8 − 1 =
19 + 4 =	7 − 5 =

16+1=	9+1=	51−0=
75+1=	39+0=	23+1=
47+0=	18+1=	34+1=

15 + ☐ =

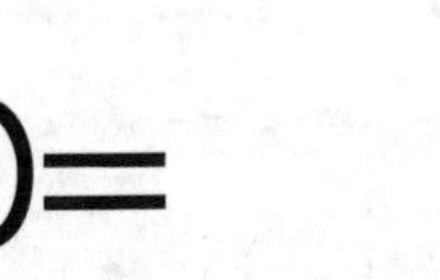

☐ −11 =

Name ____________________

4 + 5 = □

6 + □ = 10

 =

 =

26

14

□ + □ = 71

□ + □ = 24

 ____________________

$\square$

$\square$

50 8 30

$19+1=$ $3+1=$ $21+0=$

$7+1=$ $34+1=$ $18+1=$

$\square-5=$ $\square-6=$

$35+\square=$ $27+\square=$

Name ______________________________

$25 + \square =$

TTTT

=

=

=

$4 + \square = 7$

$13 + 4 = \square$

$19 + \square = 24$

24

•

45

•

□ + □ = 64 | □ + □ = 57

____

□
TTTTT
TTTT

□
TTTT
|||||||

□
||||||||

□ − 4 =
|||||||||||

19 + □ =
|||||

8 − 3 =

7 + 3 =

9+1=

17+1=

23+1=

14−0=

35+1=

19+0=

Name ____________________

$35 + \square =$

TTTT

 $5 + \square = 9$

$7 - 6 = \square$

$26 + 3 = \square$

$8 + \square = 13$

. 34

. 15

$60 + 8 = \square$

$\square + \square = 29$

$50 + 1 = \square$

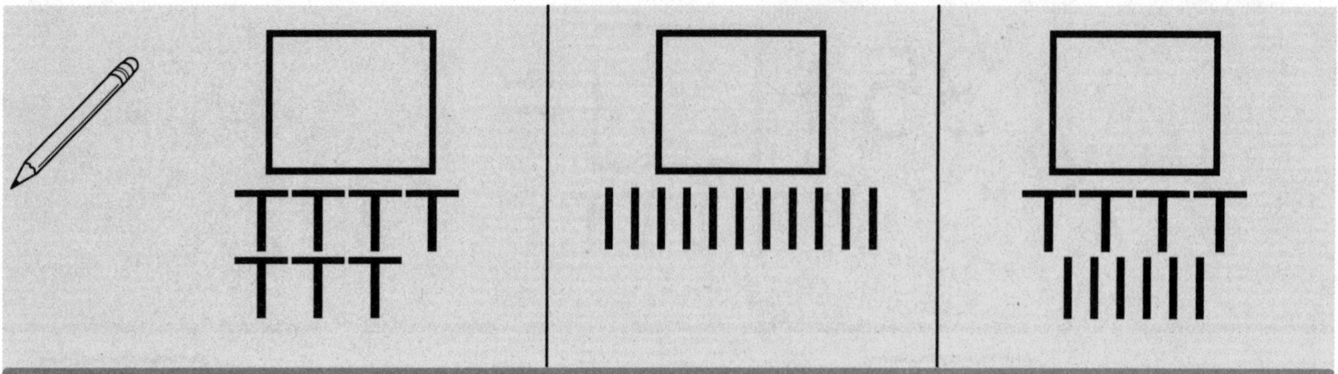

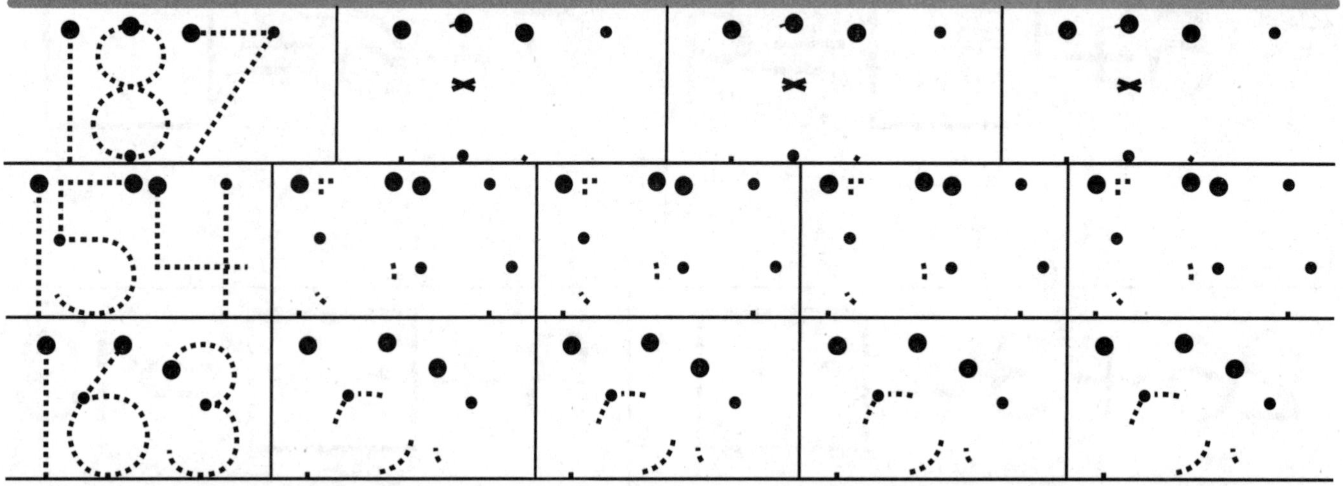

 =

 =

 =

$19 + 0 =$ | $19 + 1 =$ | $19 - 0 =$

$\square - 5 =$

$6 - 2 = \square$

Name ______________________________

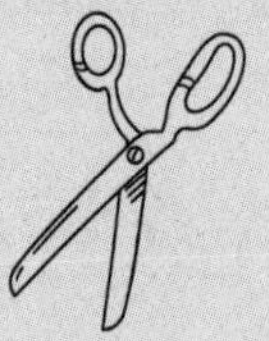

$6 - 2 = \square$

$17 + 4 = \square$ | $45 + \square = 50$

25 | $\square$ TTTTTIIII | $\square$ TIIIIII

$31 + 50 = \square$
TTTTT

$70 + 4 = \square$

$\square + \square = 91$ | $50 + 6 = \square$

 =

 =

165

139

18+1 =

76+0=

12−0=

□ −5=

26+□ =

14+3=

6−3=

Name ______________________

$36+40=\square$

TTTT

$45+20=\square$

TT

6 9

21 20

79 77

$16+\square=20$

$6-4=\square$

$37+4=\square$

____

$40 + 9 = \square$

$30 + 7 = \square$

$\square + \square = 25$

$\square$

TTTT

|||||

34

125

183

$49 + \square =$

||

$\square - 6 =$

||||||||||

$7 - 2 =$

$5 + 4 =$

Name ______________________________

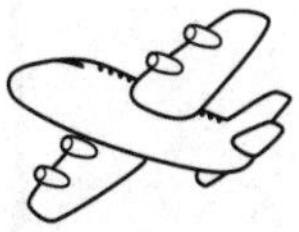 46 49 | 19 16 | 16 60

 ______________________________

14

38+4=☐

23+☐=25

8−2=☐

80+9=☐

☐+☐=34

56+40= □

34+30= □

74

38

16+1=

9+1=

40−0=

28+1=

15+1=

36+1=

□ −5=

45+ □ =

Name ____________________

85 SIDE 1

28 + □ = 33 | 5 − 4 = □

6 + 3 = □

5 + 7 = 12 ____________ | 3 + 18 = 21 ____________

 24 19 | 59 68 | 31 13

 ____________

20 + 7 = □ | □ + □ = 62

□ + □ = 58 | 80 + 5 = □

$56 + 20 = \square$

18

52

$29+1=$	$18+0=$	$45+1=$
$23-0=$	$37+1=$	$8+0=$

$38 + \square =$

$\square - 3 =$

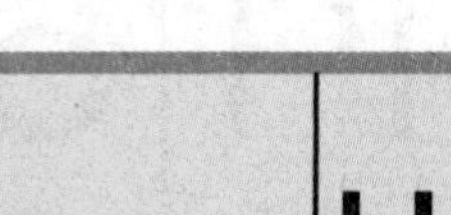

Name ____________________

$67 + 20 = \square$

$67 + 4 = \square$

$24 + 40 = \square$

☆ 12 7 | 83 91 | 29 46

$6 + 1 = \square$	$5 + 0 = \square$	$3 + 1 = \square$	$7 + 1 = \square$

$2 + 4 = 6$ ____________________

$5 + 24 = 29$ ____________________

$\square + \square = 37$

$60 + 4 = \square$

$8 - 5 = \square$

$5 + \square = 11$

$7 - 3 = \square$

$8 + \square = 12$

35

$\square$

T||||||

13

$\square$

TTT
TTT ||||

$27 - 0 =$

$49 + 1 =$

$15 + 0 =$

$\square - 6 =$

||||||||

$47 + \square =$

||||

Name ____________________

8	3	6	4	7
− 0	+ 1	+ 0	+ 1	+ 1
☐	☐	☐	☐	☐

 80 79 | 16 17 | 13 31

26 + 3 = ☐

26 + 50 = ☐

53 + 6 = ☐

53 + 30 = ☐

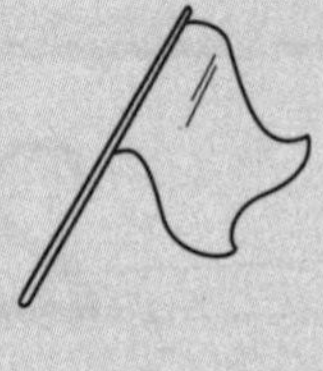 ____________________

32 + □ = 37 | 8 − 2 = □

7 + 4 = □ | 6 + □ = 9

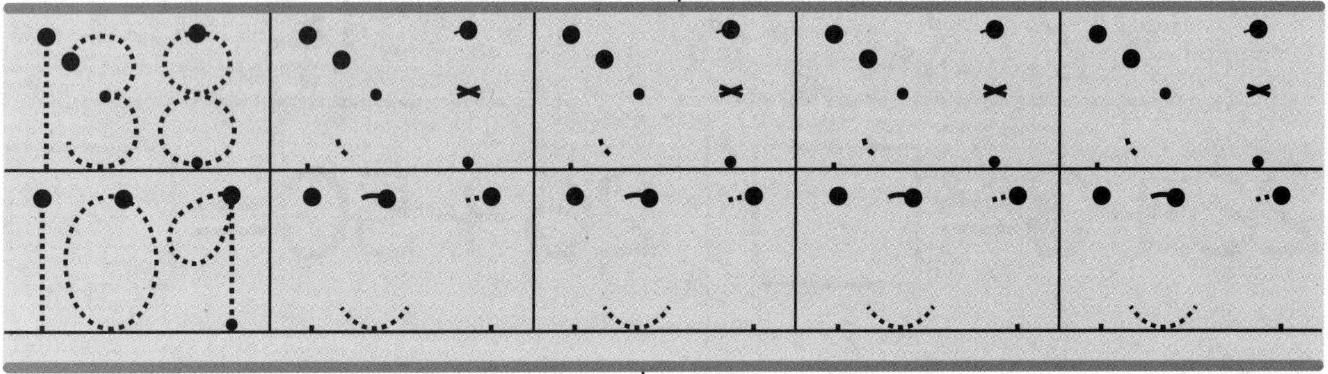

20 + 8 = □ | □ + □ = 63

40 | □ | □ | 27

T

TTT TT IIII

Name ______________________

28 + 4 =

32 + 40 =

28 + 30 =

106	99	153	150	64	46

40	10	72	19	30
+ 6	+ 8	+ 1	+ 1	+ 5
46	18	73	20	35

6 − 4 = □

8 + □ = 13

7 + 3 = □

26 + □ = 31

☐ TTTTTT	21	36	☐ TIIIII

3 − 0	4 + 1	8 + 1	7 + 0	2 + 1

40 + 6 = ☐	☐ + ☐ = 39
☐ + ☐ = 25	70 + 8 = ☐

54 + ☐ =
IIII

☐ − 7 =
IIIIIIIIII

☐ − 3 =
III

18 + ☐ =
IIII

Name ____________________

$59+40=$ | $59+4=$

$43+30=$

70	42	29	50	28
+ 6	− 0	+ 1	+ 3	+ 0
76	42	30	53	28

17 15 | 12 21 | 180 108

$65+\square=70$ | $8-3=\square$

$5+4=\square$ | $19+\square=23$

□ TTTIIIIIII	15	41

8 − 0	5 + 1	7 − 0	3 + 1	8 + 1

50 + 9 = □ | □ + □ = 41

20 + 7 = □

47 + □ =
IIIII

□ − 8 =
IIIIIIIIIII

21 + □ =
IIIIII

□ − 7 =
IIIIIIII

Name ______________________________

48	30	50
+ 1	+ 7	+ 6
☐	☐	☐

109 118	40 39	16 51

1 + 67 = ______

9 + 30 = ______

 ______________________

43 + 30 = ☐

28 + ☐ = 33

7 − 1 = ☐

25+3=□

42+□=46

35+20=□

35

□

TTT
TTT IIIIIII

12

□

TTT
TTT

□−6=

IIIIIII

□−3=

IIIIIIIIIIIII

Name ______________________

91 SIDE 1

$4 + 20 =$ ______

$1 + 15 =$ ______

 ______ ______

$$\begin{array}{r} 75 \\ +\ 1 \\ \hline \square \end{array} \quad \begin{array}{r} 60 \\ +\ 9 \\ \hline \square \end{array} \quad \begin{array}{r} 42 \\ +\ 10 \\ \hline \square \end{array} \quad \begin{array}{r} 73 \\ +\ 10 \\ \hline \square \end{array}$$

117 108	58 65	23 19

$36 + \square = 40$

$8 - 3 = \square$

$58 + 40 = \square$

7 − 4 = □

---

21 + □ = 26	48 + 3 = □

□ TTTT TTTT ǁ	12	34	□ T |||||||||

27 − 0 =	19 + 1 =	47 + 0 =

47 + □ = ||||||	□ − 7 = ||||||||||

Name ____________________

56+34=
TTTIIII

44+52=
TTTTTII

80 + 7 = ☐	25 + 10 = ☐	28 + 1 = ☐	50 + 3 = ☐

1+23= ____

9+40= ____

103 111	45 36	89 91

38 + 40 = □

7 + □ = 11

6 − 2 = □

26 + 5 = □

18 + □ = 21

45 + 20 = □

□ TTT TTT	31	13	□ TT TT IIIIIII

6 − 0 =

29 + 1 =

38 + 0 =

29 + □ =
IIIIII

□ − 4 =
IIIIIIIIIII

Name ______________________________

45+24=

TTIIIII

39+26=

TTIIIIIII

10+48=

____

28	40	53	39	19
+10	+ 6	+ 1	− 0	+ 1
☐	☐	☐	☐	☐

6−4=☐

6+☐=11

43+30=☐

8+4=☐

7 − 1 = □	28 + □ = 30

29 34	56 47	130 114	63 49

□ T||||||||	42	15	□ TTTTTT|||

□ − 2 = ||||||||||	19 + □ = ||||||

32 + □ = |||||||||	□ − 8 = ||||||||

Name ______________________________

57 + 32 =

42 + 25 =

1 + 25 = ______________

57	32	14	60	19
+ 10	− 0	+ 1	+ 5	+ 0
□	□	□	□	□

29 + 30 = □

15 + □ = 21

$8 - 2 = \square$

$9 + 6 = \square$

$41 + 20 = \square$

$36 + 5 = \square$

18 15	100 86	104 130	54 45

$\square - 10 =$

||||||||||

$54 + \square =$

||||||

$25 + \square =$

TTT

$\square - 5 =$

||||||||||||

Name ______________________________

95 SIDE 1

$31 + 22 =$

$28 + 52 =$

$5 + 60 =$

____

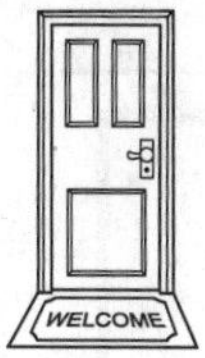

____

____

$\begin{array}{r} 25 \\ +10 \\ \hline \square \end{array}$	$\begin{array}{r} 80 \\ +\ 3 \\ \hline \square \end{array}$	$\begin{array}{r} 15 \\ +\ 1 \\ \hline \square \end{array}$	$\begin{array}{r} 37 \\ -\ 0 \\ \hline \square \end{array}$	$\begin{array}{r} 52 \\ +\ 1 \\ \hline \square \end{array}$

$47+20=\square$

$9-2=\square$

$4-1=\square$

$21+50=\square$

$6+3=\square$

$6-5=\square$

38 29 | 9 10 | 8 7

$71+\square=$

TT

$\square-1=$

||||||||||

Name ____________________

$7-3=\square$

$8+5=\square$

$4+3=\square$

$\begin{array}{r} 9 \\ +\ 3 \\ \hline \square \end{array}$	$\begin{array}{r} 7 \\ +\ 9 \\ \hline \square \end{array}$	$\begin{array}{r} 8 \\ +\ 5 \\ \hline \square \end{array}$

$33+42=$

$27+40=\square$

$6-3=\square$

$34+\square=39$

$34+20=\square$

$4-4=\square$

$18+\square=22$

49 53	5 1	117 123	8 9

Name ______________________________

$6-3=3$ ______________

$7+3=10$ ______________

$5+6=11$ ______________

$48+23=$

______________ ______________

27 + ☐ = 31	26 + 30 = ☐
34 + 3 = ☐	58 + 40 = ☐

63 58	10 2	8 7	23 16

30 + 6 ☐	12 − 0 ☐	9 + 1 ☐	60 + 8 ☐
27 + 1 ☐	42 + 0 ☐	50 + 7 ☐	39 − 0 ☐

Name ______________________________

$$\begin{array}{r} 39 \\ +\,24 \\ \hline \square \end{array}$$

____

$$\begin{array}{r} 54 \\ +\,32 \\ \hline \square \end{array}$$

____

$6-5=\square$

____

$8+5=\square$

____

$7+4=\square$

____

____

$49+20=\square$

$34+40=\square$

$26+\square=31$

$8-2=\square$

$\square-7=$

$\mid\mid\mid\mid\mid\mid\mid$

$18+\square=$

$\mid\mid\mid\mid\mid\mid\mid\mid\mid$

$28+\square=$

TTT

$\square-1=$

$\mid\mid\mid\mid\mid\mid\mid\mid\mid\mid$

50	13	10	9	30
+ 4	+ 1	− 0	+ 1	+ 8
$\square$	$\square$	$\square$	$\square$	$\square$

Name ____________________

6 + 1 = ☐

6 + 2 = ☐

17 + 1 = ☐

17 + 2 = ☐

9 + 1 = ☐

9 + 2 = ☐

3 + 1 = ☐

3 + 2 = ☐

$$\begin{array}{r} 56 \\ +\ 14 \\ \hline \square \end{array}$$ ________

TTTTT~~TTT~~II~~I~~ =

T~~TT~~IIIIIIII~~I~~ =

$7 - 5 = \square$

$15 + \square = 21$

$8 + 3 = \square$

$36 + 40 =$

$25 + 30 =$

81 19	10 6	4 5	8 9

Name ____________________

$57 - 32 =$

TTTTTIIIIIII

$12 + 1 =$	$8 + 1 =$
$12 + 2 =$	$8 + 2 =$

$$\begin{array}{r} 43 \\ +\,24 \\ \hline \square \end{array}$$

____

TTŦIIIIIǂ =

TTTTŦŦIIǂ =

36 + ☐ = 40

7 − 6 = ☐

6 + 5 = ☐

137 119	46 52	38 42	67 59

$$\begin{array}{r}40\\+\ 6\\\hline\square\end{array}\qquad\begin{array}{r}17\\+\ 1\\\hline\square\end{array}\qquad\begin{array}{r}25\\-\ 0\\\hline\square\end{array}\qquad\begin{array}{r}80\\+\ 3\\\hline\square\end{array}\qquad\begin{array}{r}9\\+\ 1\\\hline\square\end{array}$$

Name ______________________________

TTTTT
TT||| =

TTT
||||||||| =

7 + ☐ = 11

39 + 3 = ☐

8 − 6 = ☐

65 − 24 =
TTT
TTT |||||

26 + 1 =

9 + 1 =

13 + 1 =

26 + 2 =

9 + 2 =

13 + 2 =

$$\begin{array}{r} 56 \\ +\ 34 \\ \hline \square \end{array}$$

$$\begin{array}{r} 21 \\ +\ 42 \\ \hline \square \end{array}$$

$76 + 20 = \square$

$41 + 5 = \square$

$45 + 56 =$

TTTTT

IIIIII

$$\begin{array}{r} 7 \\ +\ 0 \\ \hline \square \end{array}$$

$$\begin{array}{r} 40 \\ +\ 6 \\ \hline \square \end{array}$$

$$\begin{array}{r} 9 \\ +\ 1 \\ \hline \square \end{array}$$

$$\begin{array}{r} 38 \\ -\ 0 \\ \hline \square \end{array}$$

$$\begin{array}{r} 90 \\ +\ 4 \\ \hline \square \end{array}$$

Name ____________________

$$\begin{array}{r} 53 \\ +34 \\ \hline \square \end{array}$$

$29 + 41 = \square$

$$\begin{array}{r} 9 \\ +\ 2 \\ \hline \square \end{array} \qquad \begin{array}{r} 13 \\ +\ 2 \\ \hline \square \end{array} \qquad \begin{array}{r} 8 \\ +\ 2 \\ \hline \square \end{array} \qquad \begin{array}{r} 16 \\ +\ 2 \\ \hline \square \end{array}$$

$57 - 17 =$

TTTTT
|||||||

$34 - 23 =$

$62 + \square = 67$

$9 + 5 = \square$

$7 - 4 = \square$

74 68	10 1	4 5	32 27

65+32= TTTII	28+43= TTTTIIII

□−5= |||||	□−2= |||||||||||

19−0=	20+4=	6+0=
14+1=	9+1=	70+8=
30+6=	45−0=	27+1=

Name ______________________________

$27+35=\square$

$$\begin{array}{r} 81 \\ +\ 12 \\ \hline \square \end{array}$$

$$\begin{array}{r} 12 \\ +\ 2 \\ \hline \square \end{array}$$

$$\begin{array}{r} 28 \\ +\ 2 \\ \hline \square \end{array}$$

$$\begin{array}{r} 9 \\ +\ 2 \\ \hline \square \end{array}$$

$$\begin{array}{r} 5 \\ +\ 2 \\ \hline \square \end{array}$$

$56-52=$

TTTTT

||||||

$44-14=$

$74+10=\square$

$46+10=\square$

______________________ ______________________

174 190 | 7 6 | 2 1 | 116 132

49+50=
TTTTT

23+30=
TTT

□−5=
TT IIIIIII

□−3=
T IIIII

57−0=	30+5=	18+0=
36+1=	19+1=	60+7=
50+2=	81+0=	22+1=

7+□=
III

3+□=
IIIII

Name ________________________________

$24 + 26 = \square$

$$\begin{array}{r} 36 \\ +\ 13 \\ \hline \square \end{array}$$

$$\begin{array}{r} 20 \\ +\ 7 \\ \hline \square \end{array} \quad \begin{array}{r} 25 \\ +10 \\ \hline \square \end{array} \quad \begin{array}{r} 80 \\ +\ 3 \\ \hline \square \end{array} \quad \begin{array}{r} 39 \\ +10 \\ \hline \square \end{array} \quad \begin{array}{r} 54 \\ +10 \\ \hline \square \end{array}$$

$6 - 5 = \square$

$8 + \square = 14$

________________________________ ________________________________

$$\begin{array}{r} 36 \\ -\ 13 \\ \hline \square \end{array}$$ TTTIIIIII

$$\begin{array}{r} 53 \\ -22 \\ \hline \square \end{array}$$ ________________________________

60 49	108 131	7 6	3 1

39+40= TTTT	72+20= TT
□−2= TTTIIIII	□−3= TTIIIIIII

11+1=	97−0=	80+6=
25+0=	29+1=	30+9=
41+1=	16+1=	8+1=

Name ____________________

$$\begin{array}{r} 75 \\ +\ 16 \\ \hline \square \end{array}$$

$$42 + 34 = \square$$

$$\begin{array}{r} 30 \\ +\ \ 8 \\ \hline \square \end{array}$$

$$\begin{array}{r} 83 \\ +10 \\ \hline \square \end{array}$$

$$\begin{array}{r} 70 \\ +\ \ 2 \\ \hline \square \end{array}$$

$$\begin{array}{r} 72 \\ +10 \\ \hline \square \end{array}$$

$$\begin{array}{r} 45 \\ -42 \\ \hline \square \end{array}$$ ____________________

$24+2=$	$19+2=$	$8+2=$

$65+20=\square$

$5-\square=1$

$34+\square=40$

$7+4=\square$

$8-5=\square$

$7-\square=1$

$\square-4=$

TTTIIIII

$\square-0=$

TTIIIIIII

Name ______________________________

$$\begin{array}{r} 34 \\ -\ 14 \\ \hline \square \end{array}$$

$27+2=$	$18+2=$	$9+2=$

$$\begin{array}{r} 73 \\ +\ 25 \\ \hline \square \end{array}$$

$38+42=\square$

37+30= ☐

8+ ☐ =12

6−2= ☐

49+ ☐ =51

74 82	1 10	3 1	112 109

☐ −3=

TTTT
IIIII

☐ −4=

TTTIIII

Name ____________________

6 + 40 =	2 + 13 =
1 + 27 =	3 + 70 =

$$\begin{array}{r} 45 \\ -34 \\ \hline \square \end{array} \qquad \begin{array}{r} 27 \\ +24 \\ \hline \square \end{array}$$

$$35 - 12 = \square$$

64 57	3 4	140 116	8 3

52+40=□

39+□=45

8−2=□

7+□=10

4−3=□

21+30=□

□−2=

T IIIIIIII

□−6=

TTT

IIIIIII

Name ______________________

ŦŦŦ|||ǂ =   TTTTŦŦ⧺ =

$\begin{array}{r} 27 \\ -\ 12 \\ \hline \square \end{array}$	$\begin{array}{r} 46 \\ +\ 14 \\ \hline \square \end{array}$

$45 - 23 = \square$

$\begin{array}{r} 2 \\ +\ 8 \\ \hline \square \end{array}$	$\begin{array}{r} 1 \\ +35 \\ \hline \square \end{array}$	$\begin{array}{r} 3 \\ +60 \\ \hline \square \end{array}$	$\begin{array}{r} 2 \\ +23 \\ \hline \square \end{array}$

5 − 1 =	7 − 1 =
10 − 1 =	3 − 1 =

$18 + \square = 21$

$7 - 4 = \square$

$46 + 30 = \square$

$50 + \square = 56$

Name ______________________________

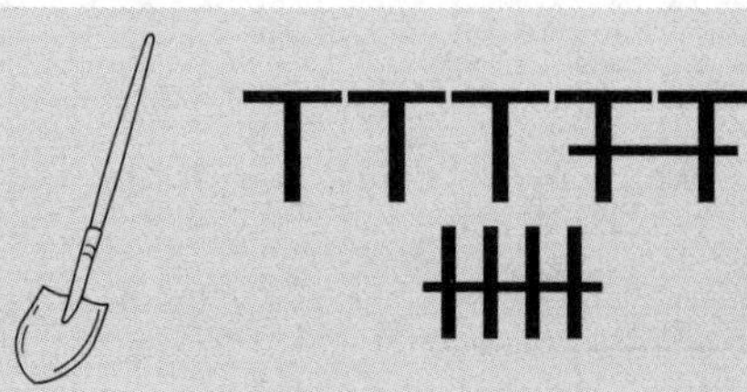 =

 =

$$\begin{array}{r} 35 \\ -\ 15 \\ \hline \square \end{array}$$

$$\begin{array}{r} 48 \\ +\ 32 \\ \hline \square \end{array}$$

$$\begin{array}{r} 10 \\ +\ 25 \\ \hline \square \end{array}$$

$$\begin{array}{r} 2 \\ +\ \ 4 \\ \hline \square \end{array}$$

$$\begin{array}{r} 10 \\ +\ 47 \\ \hline \square \end{array}$$

$$\begin{array}{r} 2 \\ +\ 27 \\ \hline \square \end{array}$$

3 − 1 =

6 − 1 =

4 − 1 =

7 − 1 =

 ______________________

113 109	4 5	70 68	4 1

$8 - 5 = \square$

$38 + 40 = \square$

$9 + \square = 14$

$36 + 4 = \square$

$\square - 3 =$

|||

$25 + \square =$

|||||

Name ______________________

13+10= ☐	18+10= ☐
11+10= ☐	14+10= ☐

$$\begin{array}{r} 43 \\ -\ 23 \\ \hline \square \end{array} \qquad \begin{array}{r} 27 \\ +\ 34 \\ \hline \square \end{array}$$

45+25= ☐

10+24= ☐	2+18= ☐
2+5= ☐	10+15= ☐

 ______________________

10	7	7	4	6	9
$-\ 0$	$+\ 1$	$-\ 1$	$-\ 1$	$+\ 1$	$-\ 1$
$\square$	$\square$	$\square$	$\square$	$\square$	$\square$

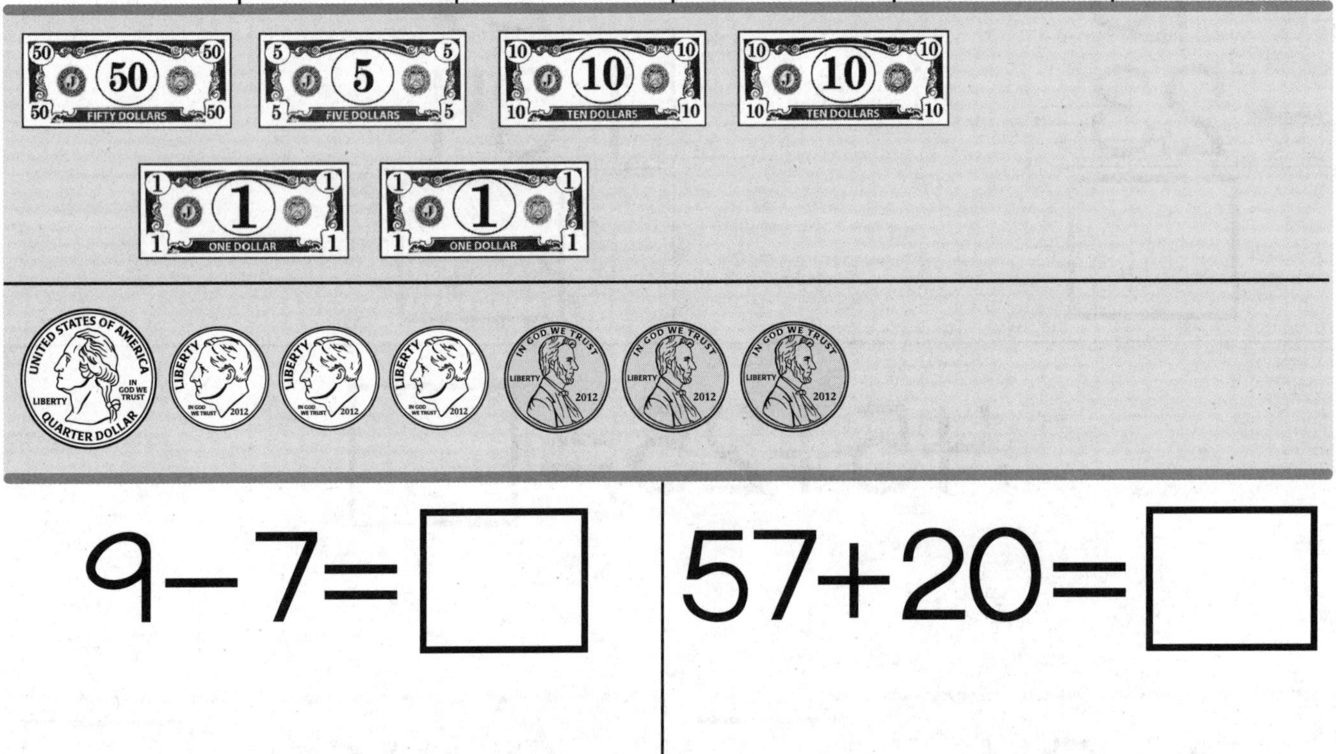

$9 - 7 = \square$

$57 + 20 = \square$

$29 + \square = 35$

$8 - 4 = \square$

Name ______________________

12 + 10 = ☐ 17 + 10 = ☐

15 + 10 = ☐ 14 + 10 = ☐

$$\begin{array}{r} 52 \\ +32 \\ \hline \square \end{array} \qquad \begin{array}{r} 54 \\ -30 \\ \hline \square \end{array}$$

$$\begin{array}{r} \square \\ +\square \\ \hline 40 \end{array} \qquad \begin{array}{r} \square \\ +\square \\ \hline 73 \end{array} \qquad \begin{array}{r} \square \\ +\square \\ \hline 50 \end{array} \qquad \begin{array}{r} \square \\ +\square \\ \hline 31 \end{array}$$

8 − 1 = 9 + 1 =

4 + 1 = 6 − 1 = 10 − 1 =

10 + 2 =

30 + 5 =

20 + 7 =

10 + 3 =

58 63	7 6	4 1	115 121

65 + 30 = ☐

9 − 2 = ☐

28 + ☐ = 32

4 − 3 = ☐

$17 + 10 = \square$	$43 + 10 = \square$	$11 + 10 = \square$	$15 + 10 = \square$	$38 + 10 = \square$	$16 + 10 = \square$

$9 + 42 = \square$	$53 - 13 = \square$

$10+3=$	$20+6=$
$70+8=$	$10+5=$

$9 - 1 = \square$	$10 + 1 = \square$	$5 - 1 = \square$	$8 + 1 = \square$	$2 - 1 = \square$	$10 - 1 = \square$

$\square + \square = 84$	$\square + \square = 60$	$\square + \square = 59$

____

____

$8-3=\square$

$28+40=\square$

Name ______________________________

8+2=	4−1=	9+1=
7+2=	10−1=	5+2=

13+41=	26−22=

$$\begin{array}{r} 4 \\ +\ 36 \\ \hline \end{array}$$

 □ + □ = 90

□ + □ = 13	□ + □ = 58

____

$$\begin{array}{r} 12 \\ +10 \\ \hline \square \end{array} \qquad \begin{array}{r} 47 \\ +10 \\ \hline \square \end{array} \qquad \begin{array}{r} 51 \\ +10 \\ \hline \square \end{array} \qquad \begin{array}{r} 15 \\ +10 \\ \hline \square \end{array}$$

$5 - \square = 1$

____

$35 + 20 = \square$

____

$8 - 4 = \square$

____

$19 + \square = 24$

____

Name ____________________

$54 - 23 =$

$57 + 23 =$

$$\begin{array}{r} 32 \\ -\ 12 \\ \hline \end{array}$$

2+10=	7−1=	63+2=
7+1=	2−1=	17+2=
63+10=	3−1=	17+10=

$$\begin{array}{r} \square \\ +\ \square \\ \hline 12 \end{array} \qquad \begin{array}{r} \square \\ +\ \square \\ \hline 20 \end{array} \qquad \begin{array}{r} \square \\ +\ \square \\ \hline 69 \end{array} \qquad \begin{array}{r} \square \\ +\ \square \\ \hline 10 \end{array}$$

$32+40=\square$

$7-6=\square$

$27+\square=32$

$6-5=\square$

Name ______________________

$$\begin{array}{r} 44 \\ -12 \\ \hline \square \end{array}$$

$$\begin{array}{r} 5 \\ +26 \\ \hline \square \end{array}$$

$$36-30=\square$$

18+10=	8−1=	35+2=
5−1=	8+2=	10−1=
58+10=	18+1=	5+10=

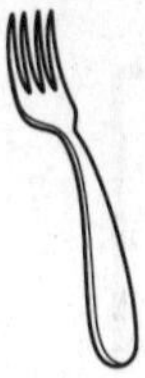

$$\begin{array}{r} \square \\ +\ \square \\ \hline 48 \end{array} \quad \begin{array}{r} \square \\ +\ \square \\ \hline 12 \end{array} \quad \begin{array}{r} \square \\ +\ \square \\ \hline 80 \end{array} \quad \begin{array}{r} \square \\ +\ \square \\ \hline 54 \end{array}$$

$5-3=\square$

$38+20=\square$

$76+\square=80$

$9-4=\square$

Name ____________________

$6 + \square = 10$

$7 + \square = 10$

$8 + \square = 10$

$9 + \square = 10$

7 − 1 =

|||||||

7 − 2 =

|||||||

7 − 3 =

|||||||

7 − 4 =

|||||||

3 4 ___ 6 ___ 8 ___ 10 ___

7+10=	7−1=	7+2=
30+7=	13+1=	17−0=
8+2=	28+10=	8−1=

TIIIII=	TIIIIII=

$$\begin{array}{r} 6 \\ +35 \\ \hline \square \end{array} \qquad \begin{array}{r} 56 \\ -23 \\ \hline \square \end{array}$$

27+13= □

$$\begin{array}{r} \square \\ +\ \square \\ \hline 13 \end{array} \quad \begin{array}{r} \square \\ +\ \square \\ \hline 70 \end{array} \quad \begin{array}{r} \square \\ +\ \square \\ \hline 50 \end{array} \quad \begin{array}{r} \square \\ +\ \square \\ \hline 89 \end{array}$$

Name ______________________

5−0=	7+2=	6−1=
4+1=	10+1=	9+2=
2+0=	7−1=	8+1=
5+2=	8−1=	3+2=

4+☐=10	5+☐=10
______	______
6+☐=10	7+☐=10
______	______

6	7	10	9	8	11
☐	☐	☐	☐	☐	☐

$6-1=$

IIIIII

$6-2=$

IIIIII

$6-3=$

IIIIII

$6-4=$

IIIIII

______ ______

 TIIIIIIII =

= TIIII TII =

$$\begin{array}{r} 8 \\ +24 \\ \hline \square \end{array} \qquad \begin{array}{r} 26 \\ -12 \\ \hline \square \end{array}$$

$\square + \square = 95$ $\square + \square = 19$

Name ______________________________

$10-4=$ ||||||||||	$10-5=$ ||||||||||
$10-6=$ ||||||||||	$10-7=$ ||||||||||

$7+2=$	$40+7=$	$13+10=$
$3-1=$	$3+1=$	$8+2=$
$20+8=$	$9-0=$	$8-1=$
$4+2=$	$4-1=$	$9+1=$

14 17 16 15 18 19 20

_______________ _______________

P R R T P S P P R T P

P= R= S= T=

	1	2	3	4	5	6
Letter						

6 + □ = 9

3 + □ = 7

2 + □ = 8

1 + □ = 4

37
+ □ TIIII

□ TTTTTIIIII
− 35

Name ____________________

9+0=	9−1=	80+4=
5−1=	5+2=	17+10=

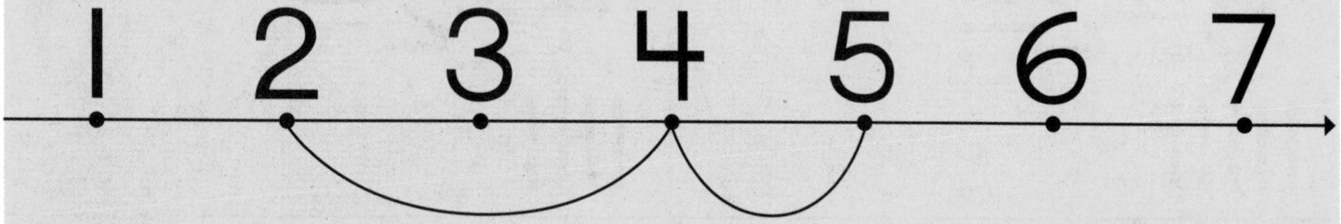

2 ____	4+ ____
4− ____	5 ____

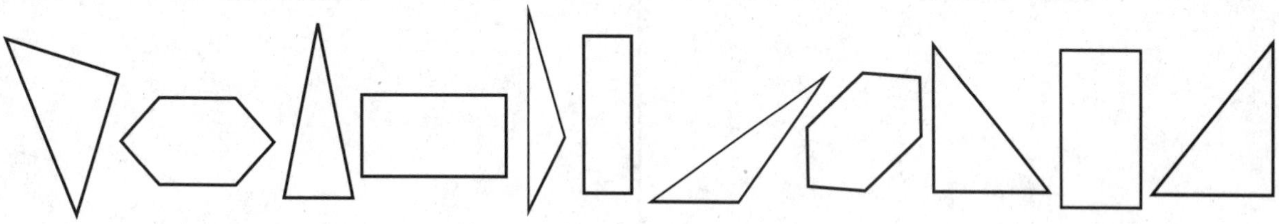

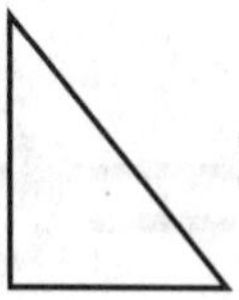 =   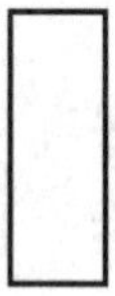 =    =

	1	2	3	4	5	6
shapes						

12 13 41 15 11 17 15 91 20

5−2= IIIII	4−2= IIII
3−2= III	2−2= II

4+ ☐ =9

2+ ☐ =7

1+ ☐ =5

3+ ☐ =5

Name ______________________________

13+10=	10+4=	7+10=
40+9=	36+10=	30+6=

1 2 13 4 7 6 8 9

4−4= IIII	7−2= IIIIIII
8−3= IIIIIIII	5−4= IIIII

5 6 7 8 9

6 ____	8+ ____
8− ____	9 ____

6 + □ = 10

3 + □ = 7

2 + □ = 5

2 + □ = 7

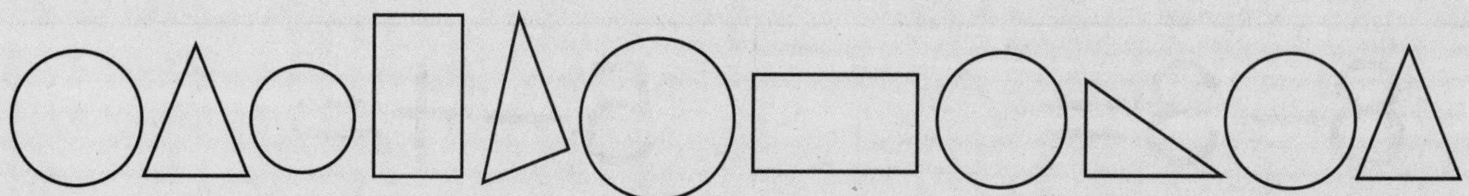

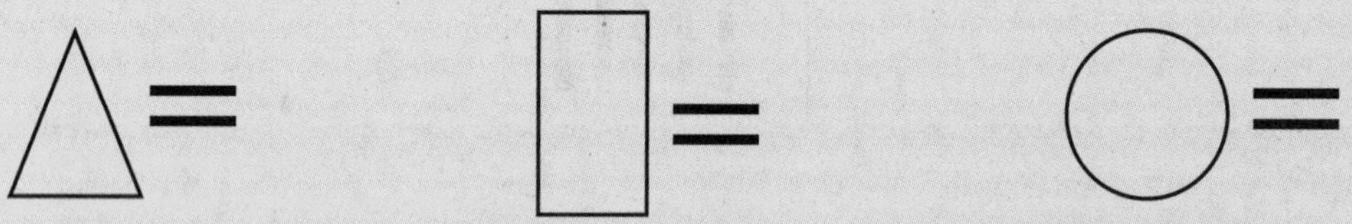

	1	2	3	4	5	6
shapes	△ ▭ ○	△ ▭ ○	△ ▭ ○	△ ▭ ○	△ ▭ ○	△ ▭ ○

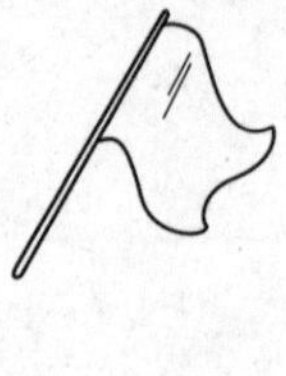

Level A Correlation to Grade K

Common Core State Standards for Mathematics

Counting and Cardinality (K.CC)

Know number names and the count sequence.

1. Count to 100 by ones and by tens.

Lessons	66–75, 80, 81

Counting and Cardinality (K.CC)

Know number names and the count sequence.

3. Write numbers from 0 to 20. Represent a number of objects with a written numeral 0–20 (with 0 representing a count of no objects).

Lessons	1–100

Counting and Cardinality (K.CC)

Count to tell the number of objects.

4. Understand the relationship between numbers and quantities; connect counting to cardinality.
   a. When counting objects, say the number names in the standard order, pairing each object with one and only one number name and each number name with one and only one object.
   b. Understand that the last number name said tells the number of objects counted. The number of objects is the same regardless of their arrangement or the order in which they were counted.
   c. Understand that each successive number name refers to a quantity that is one larger.

Lessons	15–120

Counting and Cardinality (K.CC)

Count to tell the number of objects.

5. Count to answer "how many?" questions about as many as 20 things arranged in a line, a rectangular array, or a circle, or as many as 10 things in a scattered configuration; given a number from 1–20, count out that many objects.

Lessons	15–120

Counting and Cardinality (K.CC)

Compare numbers.

6. Identify whether the number of objects in one group is greater than, less than, or equal to the number of objects in another group, e.g., by using matching and counting strategies.

Lessons	24–33, 36, 44–55, 59, 60

Operations and Algebraic Thinking (K.OA)

Understand addition as putting together and adding to, and understand subtraction as taking apart and taking from.

1. Represent addition and subtraction with objects, fingers, mental images, drawings, sounds (e.g., claps), acting out situations, verbal explanations, expressions, or equations.

Lessons	26–120

Operations and Algebraic Thinking (K.OA)

Understand addition as putting together and adding to, and understand subtraction as taking apart and taking from.

2. Solve addition and subtraction word problems, and add and subtract within 10, e.g., by using objects or drawings to represent the problem.

Lessons	36–120

Operations and Algebraic Thinking (K.OA)

Understand addition as putting together and adding to, and understand subtraction as taking apart and taking from.

3. Decompose numbers less than or equal to 10 into pairs in more than one way, e.g., by using objects or drawings, and record each decomposition by a drawing or equation (e.g., 5 = 2 + 3 and 5 = 4 + 1).

Lessons	76–89, 111–120

Operations and Algebraic Thinking (K.OA)

Understand addition as putting together and adding to, and understand subtraction as taking apart and taking from.

4. For any number from 1 to 9, find the number that makes 10 when added to the given number, e.g., by using objects or drawings, and record the answer with a drawing or equation.

Lessons	116–120

Operations and Algebraic Thinking (K.OA)

Understand addition as putting together and adding to, and understand subtraction as taking apart and taking from.

5. Fluently add and subtract within 5.

Lessons	56–120

Number and Operations in Base Ten (K.NBT)

Work with numbers 11–19 to gain foundations for place value.

1. Compose and decompose numbers from 11 to 19 into ten ones and some further ones, e.g., by using objects or drawings, and record each composition or decomposition by a drawing or equation (e.g., 18 = 10 + 8); understand that these numbers are composed of ten ones and one, two, three, four, five, six, seven, eight, or nine ones.

Lessons	66–93, 102–117

Measurement and Data (K.MD)

Describe and compare measurable attributes.

2. Directly compare two objects with a measurable attribute in common, to see which object has "more of"/"less of" the attribute, and describe the difference. *For example, directly compare the heights of two children and describe one child as taller/shorter.*

Measurement and Data (K.MD)

Classify objects and count the number of objects in each category.

3. Classify objects into given categories; count the number of objects in each category and sort the categories by count.

Lessons	118–120

Geometry (K.G)

Identify and describe shapes (squares, circles, triangles, rectangles, hexagons, cubes, cones, cylinders, and spheres).

1. Describe objects in the environment using names of shapes, and describe the relative positions of these objects using terms such as *above, below, beside, in front of, behind,* and *next to.*

Geometry (K.G)

Identify and describe shapes (squares, circles, triangles, rectangles, hexagons, cubes, cones, cylinders, and spheres).

2. Correctly name shapes regardless of their orientations or overall size.

Geometry (K.G)

Analyze, compare, create, and compose shapes.

5. Model shapes in the world by building shapes from components (e.g., sticks and clay balls) and drawing shapes.

Lessons	eLessons 100–101

Geometry (K.G)

Analyze, compare, create, and compose shapes.

6. Compose simple shapes to form larger shapes. *For example, "Can you join these two triangles with full sides touching to make a rectangle?"*

Lessons	eLessons 102–104

Standards for Mathematical Practice

Connecting Math Concepts addresses all of the Standards for Mathematical Practice throughout the program. What follows are examples of how individual standards are addressed in this level.

1. Make sense of problems and persevere in solving them.

Symbol Identification and Symbol Writing (Lessons 1–120): Students learn to identify and write numbers and symbols, such as plus, minus, and equals, extending that information to first translating and then solving word problems.

2. Reason abstractly and quantitatively.

Numbers and Counters (Lessons 1–120): Students learn to represent numbers and quantities with counters and apply that knowledge to adding, subtracting, and solving word problems. They learn to add and subtract using Ts and lines, where a T represents 10 and a line represents 1.

3. Construct viable arguments and critique the reasoning of others.

Shapes and 3-D Objects (Lessons 85–120): Starting in Lesson 116, students learn the difference between 2-dimensional and 3-dimensional objects, acquiring the knowledge to construct an argument for whether an object is 2-dimensional or 3-dimensional.

4. Model with mathematics.

Word Problems (Lessons 25–120): Students write and solve equations that describe situations in word problems.

5. Use appropriate tools strategically.

Throughout the program (Lessons 1–120) students use pencils and their workbooks to complete only the work that the teacher instructs them to complete.

6. Attend to precision.

Counting (Lessons 1–120): Students count objects (e.g., Ts, lines, coins) to an exact number without stopping short or going beyond. This practice is the foundation for the addition and subtraction strategies that students later learn.

7. Look for and make use of structure.

Commutative Property (Lessons 77–120): Students learn that they can "turn around" addition problems so that when they know $3 + 2 = 5$, they also know that $2 + 3 = 5$.

8. Look for and express regularity in repeated reasoning.

Column Problems (Lessons 85–120): Students add and subtract single-digit numbers by counting and representing numbers with counters (e.g., Ts and lines). They apply a parallel strategy to add and subtract two 2-digit numbers. What works for adding and subtracting basic facts also works for bigger numbers.